Looking at the Sky

Characters

Red Group

Blue Group

Purple Group

All

Setting Outside, day and night

by Jeffrey B. Fuerst

My Picture Words

clouds

moon

stars

sun

My Sight Words

I	look
see	the
up	you

Look up!

Can you see the ☀ ? 
sun

Yes. I see the ☀. 
sun

5

Look up!

Can you see the 🌥 ? 
clouds

Yes. I see the 🌥 . 
clouds

7

Look up!

Can you see the 🌙 ?
moon

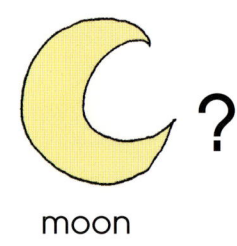

Yes. I see the 🌙 .
moon

9

Look up!

Can you see the ⭐⭐?
stars

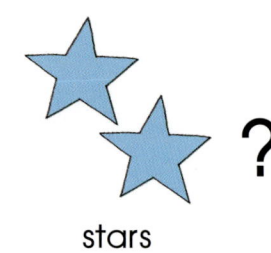

Yes. I see the ⭐⭐.
stars

11

Look up!

The End